Charles Denison

Exercise and Food for Pulmonary Invalids

Charles Denison

Exercise and Food for Pulmonary Invalids

ISBN/EAN: 9783744646574

Printed in Europe, USA, Canada, Australia, Japan

Cover: Foto ©berggeist007 / pixelio.de

More available books at **www.hansebooks.com**

EXERCISE AND FOOD

FOR

PULMONARY INVALIDS

. BY

CHARLES DENISON, A.M., M.D.

DENVER, COLORADO

PROFESSOR OF DISEASES OF THE CHEST AND OF CLIMATOLOGY, UNI-
VERSITY OF DENVER; EX-PRESIDENT AMERICAN CLIMATOLOG-
ICAL ASSOCIATION; AUTHOR OF "THE ROCKY MOUNTAIN
HEALTH RESORTS;" "THE CLIMATES OF THE
UNITED STATES, IN COLORS;" "THE
PREFERABLE CLIMATE FOR
CONSUMPTION," AND
OTHER PAPERS

.

DENVER
THE CHAIN & HARDY CO.
1895

THESE two essays have elicited considerable commendation from brother physicians. My patients, too, have seemed to need just such an opportunity as here presented, to study into the important aids to cure. The necessity of much talking on the part of the physician is avoided, and it is well for the invalid to transfer so much of the advice given from the consulting office to his own home and kitchen. I have, therefore, been led to believe this combined issue would meet a popular demand.

The "rest cure" is much in vogue just at present. There are also many writers who even decry (though rather vaguely) the exercise as well as the high-altitude treatment of consumption. I wish, therefore, to say that I feel the responsibility of thus defending physical development in disease. These pages are written advisedly, and are backed by a certain right on my part to speak with some authority on the subject of *exercise*, as well as of *food, for invalids*. In college I was an instructor in gymnastics, and afterward having neglected exercise I came to Colorado an invalid. So personal experience and the advising of over 3,000 invalids in Colorado seem to warrant some positiveness on my part.

The dedication of this little volume is to the many living examples of the benefits furnished by the observance of the principles here imperfectly expounded.

CHARLES DENISON.

DENVER, *January*, 1895.

EXERCISE

FOR

PULMONARY INVALIDS.

Revised Edition

EXERCISE

FOR

PULMONARY INVALIDS.

I BELIEVE it was Artemus Ward who proposed to get up a new life insurance company to insure people only while they were "in bed." He had looked into the matter extensively, and found that almost everybody died in bed, so he concluded that was a dangerous place.

As usual with this genius, there was some wisdom in his wit, for sedentary life and respiratory inactivity are undoubtedly largely explanatory of the preference tuberculosis shows for the pulmonary organs. Why is it that consumption is so often "of the lungs," and not — at first, at least — of the rest of the body?

1. Is it because of the non-use of certain portions of the lungs, the apices (tops) particularly, in chest-

Read before the Congress of Medico-Climatology of the World's Fair Congress Auxiliary, June 1, 1893, and printed in response to a resolution of the same.

bent clerks and flat-chested people of sedentary occupations?

2. Does consumption come by preference in the lungs because the circulation of blood there, in the capillaries, is normally five times as rapid as in the periphery of the body, and therefore any stagnation of this movement, made proportionally worse by any lessened excretion elsewhere as in the skin, liver, bowels, and kidneys, must result in unnatural conditions inimical to the health of those delicate parts? Is it true that imperfect circulation has so much to do with the pretubercular state as the above supposition would seem to indicate?

3. Is it because catarrhal conditions in the pulmonary air cells and bronchial tubes, caused by colds, influenzas, or lung engorgements, when not thrown off by good lung ventilation, furnish specially favorable culture beds for the multiplication of tuberculosis germs?

4. Is it because the ordinary breathings of sedentary people only remove a tenth at a time of the air the lungs contain, and those portions of the lung apices and periphery, farthest from the large tubes, are so little disturbed that they become vitiated and retroactive in the self-poisoning process through which the individual passes?

5. Is it because the bacillus of tubercle needs some such vitiated climate, be it the stagnant, imprisoned air or the chemically changed secretions,

in order to multiply most prolifically, until it is even possible to find, as we have done, a thousand germs to a small fly-speck of sputum?

6. Is it, as I have conceived it possible, because bacillary growth and reproduction are sooner inaugurated by the process of alternating rarefaction and condensation in these bacilli-laden air cells, a process synchronous with and due to the respiratory acts — inspiration and expiration? That is, that the septic juices which precede the germs, or the spores of the germs themselves, are churned into a quicker existence, perhaps somewhat as butter is made by the agitation of cream?

7. Or is it, as for all that I know I may be responsible for supposing, that the pulmonary air cells are the natural dumping-off places, where through the agency of the leucocytes, nature intends to do the cleansing of the blood of such impurities as these invading bacilli? These leucocytes, or white blood corpuscles, are known to be able to carry the bacilli of tuberculosis, embraced within their delicate bodies, through the attenuated alveolar (air cell) walls. Therefore, as the contents of these air cells in active tuberculosis are seen by the microscope to be made up almost wholly of these two, leucocytes and germs, is it not reasonable to infer that these carriers, and scavengers of the body brought the bacilli there?

8. Or finally, does consumption by preference come in the lungs because this course is natural in

the delicate meshes of the pulmonary structure, so wonderfully endowed with capillary vessels and glandular tissue? Because directly where the interchange from heavy venous to light arterial blood takes place (the process of hæmatosis—blood making—in intervascular and intertubular spaces) nature finds the best places to arrest the invading bacilli by her own protecting agencies? Among these agencies are the leucocytes, whose "policeman function" enables them to arrest, and then to build fibroid (new tissue) encasements around, the invading enemy.

Nature, ever bountiful in her provisions and generously replying to the unsatisfied demands of continued irritation, undoubtedly overdoes this pasting in and shutting off these bacilli until whole lobules and lobes become involved in this extradition process. Thus those suburban regions of the lungs, naturally suited for the purpose, become the burying ground of these conquered (?) bacilli. This is in accord with post-mortems, where the lungs' apices, periphery, and peribronchial spaces are found to be crowded with these bacilli-filled prison vaults and encapsulated products. This process is often seen to be advanced even to the total annihilation of the respiratory functions ; so that the affected portion of lung is, even during the life of the person, only a little better than dead.

Settle as you will this question why tuberculosis attacks by preference the lungs, whether you answer in the affirmative one or all of these eight questions, or consider these explanations as variously applicable to different cases, you must come to the decision that it is *natural elimination* which is interfered with, and it is *healthful respiration* which is wanted. It is *action* as opposed to *stagnation*. It is prevention of any further mischief which is the cure. It is cleansing as opposed to bacillary life. The time and opportunity for *rest*, as in the repair of disabled machinery, is not granted to the respiratory organs, especially not now when they are charged with a double function, that of supplying oxygen to the blood and of purifying the system of these bacillary and septic foes. The germs, when not encapsulated and so prevented from growth or restrained, multiply in geometrical progression, perhaps quicker when harbored in the warm and constantly moving lung than when quiet elsewhere.

If it were not for this, to the individual, unconscious opposition on the part of nature, humanity would be shortly blotted out from the face of the globe by this infinitesimal enemy which Robert Koch discovered.

In this fight of life against lives, it is one against billions, maybe ; but intelligence is, or ought to be, on the side of *the one*. To that one, great possibilities of muscular and respiratory power are granted, indicative of healthful resistance and elimination.

It is a man's *living cell*, his own vitality — his

selective and eliminative forces — which must be
"toned up" to a natural and healthful standard.
Then bacilli cannot find a home in him. This is the
function of *exercise*, especially with reference to the
respiratory organs,— to enable one to feel that all his
powers are "up to par," that every muscle and fibre
which tend to make up a full-chested, symmetrical
man, are daily brought into play.

Let it be distinctly understood that, in recommend-
ing exercise for invalids, acute and active inflammatory
conditions of the lungs are excluded. I am referring
particularly to the after-effects of such accidents as
these acute conditions, and to the faulty states incident
to or productive of slow encroachments of pulmonary
tuberculosis. High fevers have a cause, perhaps previously undetected, and, of course, demand rest and
soothing treatment. Pneumonias and pleurisies have
to be fought by the aid of enforced rest to the respiratory organs; and pulmonary hemorrhage can sometimes be mechanically stopped. It is, indeed, "an
ill wind that blows nobody any good," and hemorrhage from the lungs has proved "a blessing in
disguise" to many a young man, including the writer
hereof, by temporarily throwing him off his feet, and
giving him an opportunity to seasonably and fully
appreciate that there was a latent demand in his
system for a change in his environment and physical
activities.

EXERCISE AS COMPARED WITH OTHER PREVENTIVE
AND REMEDIAL TREATMENTS FOR PHTHISIS.

Even before a bacillary cause of consumption was
understood, it was generally recognized by medical
men that the affected human organism must be
helped to *itself* throw off the disease. There has
been no defensible excuse given for the nonsensical
and crazy idea of sending consumptives to live down
in the natural cave in Kentucky, that they might
have the benefit (?) of an *equable temperature;* nor
of the almost equally hazardous plan of housing
patients in dark rooms without any ventilation for
fear of a *draft.* The value of an active life out of
doors, associated with generous diet, has always
been recognized by the best medical men. But the
object of foods specially suited to combat tuberculosis — such as the Salisbury meat diet, fats and
oils, with or without stimulants — is in exact
agreement with the effects of systematic and adapted
exercise. Through the former there is a generous
and proper supply to meet an unnatural waste, and
through the latter a competent circulation and a muscular tone to keep up a normal elimination. It
must be, indeed, a peculiar accident which will result
in nature's harboring tuberculosis in a *physically* perfect human being. But accidents will come and
perfection to meet them is very seldom or never
found. For this reason are we doctors! When we

come to consider remedial measures under the head
of *medicines*, we have to acknowledge that the most
of them that are of any use are of the same nature —
appropriate foods either for the wasting tissues or
for the depraved nervous force. Such are cod liver
and petroleum oils, hypophosphites, and some mine-
ral tonics. As to drugs, how unsatisfactory they are! To
be sure, we can neutralize excessive fermentation
and aid digestion. We can also aid elimination in
different ways, as, for instance, by large hot water
flushings of the bowels when indicated, as is often
the case ; by the use of mercurial and other inunc-
tions ; and by oxygen inhalations. However, these
are all substitutes for what ought rather to be accomplished by systematic hygienic care of the body with
exercise as a leading feature.

As to saturating the system with antiseptics so
thoroughly that the tubercle bacilli *may not* be
permitted to live and thrive while the human being
harboring them *may*, the thing is probably impossible.
It is to my mind doubtful if creosote, one of the
latest fads in medical lore, has any more effect upon
the invading bacillus than to neutralize its ptomaines
—the septic results of its having already existed.
This, if it were not for upsetting the digestion, is,
of course, salutary. It is better to accept the inevi-
table and to trust to the individual *cell* of the man,
strengthened by symmetrical development, to the

end that a healthful and vigorous respiratory and circulatory system be maintained.

Quite as important is the possibility of substituting exercise for certain other methods of treatment, known or believed to favor the natural encapsulation, and so quietus, of the bacilli. I refer to the hypodermic use of such combinations as the cantharidates or chlorides of different bases — as the double chloride of gold, or of iodine also (on the Shurley-Gibbs plan.) The hardening and shrinkage of tissues around tubercular infected spots in the lungs, which seem to be the temporary local effect of such methods, is somewhat like the effect of appropriate exercise, and like the results seen of exaggerated respiratory activity in high altitudes.

The importance of exercise is not lessened, but increased, when we come to consider the use of Professor Koch's tuberculin, or of either Klebs' or Hunter's modifications of it. Here, to a new healing propensity, not unlike, but in exaggeration of, that already described as due to the chlorides of gold, etc., is added a natural stimulus — an *immunity effort on the part of nature* which is the most wonderful discovery in recent medical history.

A complete understanding of this process has not been reached, but a close clinical observation of its effects along natural lines brings *exercise* to the front as an invaluable aid to this natural method of cure. This is most reasonable; for while we do not yet know the per cent. of tuberele baeilli which are killed by this process of immunization, the shrinkage of the affected lung tissues probably leaves the germs in the intervening or included spaces. These spaces must then be more or less open in the spongy lung because of the shrinkage mentioned. It is suicidal to do this only and then stop, or by forced inactivity of the respiratory apparatus to allow these germs to remain in conditions undoubtedly more favorable than before for reinfection and reproduction.

Exercise, commensurate with these specific effects, is the *sine qua non* of success. Hence, one important reason why better results have been achieved with tuberculin, tuberculocidin or antiphthisine in high altitudes than at sea level, especially more than in any hospital practice, is because life in high altitudes is an imperative and continued activity of the respiratory apparatus. There is to me a peculiar significance in the harmony of effects of altitude, hill-climbing, and tuberculin reaction upon slightly affected areas of lung tissue. The dryer, harsher, louder, and more tubular breath-sounds locally distinguished during the first few weeks' treatment are quite similar under all three of these conditions. During his recent visit to Colorado I was pleased to hear Dr. C. Theodore Williams of London say, as we were examining a patient who had been assiduously climbing hills at eight thousand feet elevation, that he had frequently noticed in the altitude cure the character of respiratory sounds heard in this case, to which I had called his attention as being similar to the tuberculin reaction sound.

All the most valuable attributes of climate in the cure of consumption — the *sunshine* and *dryness*, the *lessened atmospheric pressure* and *electric stimulation* — are in their results in perfect accord with exercise. The peculiar dry air of interior elevated sections has the same effect through a more rapid lung action favoring the abstraction of heat and moisture from the fevered and catarrhal lung; the rarefied air and electric tension are in themselves most important stimulators of respiratory gymnastics. The mountainous configuration of most high places invites to hill climbing, and the increased radiation of heat due to rarefaction and dryness is a natural incentive to, or substitute for, exercise. These agencies are to be preferred because they are natural; and besides the enforced observation of artificial rules is less imperative in high altitudes.

BREATHING EXERCISES.

The practice of inhaling medicines properly comes under the head of respiratory exercise; so much so, that I very much doubt if the atomized antiseptics, or whatever is inhaled, have had the good effect ascribed to them; but rather the effect has been due to the respiratory gymnastics necessitated in the process of inhalation. The truth is, except in the upper air tract, the medicaments have not ordinarily reached the infected portions. The fibroid process of healing, in other words the encapsulation of tubercle, which is the most prominent feature of the natural arrest or chronicity in lung tuberculosis, tends to close or shut off the affected portions from the inhaled air. In this protecting policy the whole lung is more or less implicated, and the inspired remedy usually goes with the inhaled air to the unaffected lung on the other side. For this reason a great deal of the inhalation treatment, so popular with certain specialists, is utterly useless; or by irritating sound lung areas is positively harmful. In this way the utility of the "Howe tube" is explained, which the Rev. Dr. Buckley, of *The Christian Advocate*, has done so much to popularize because of the good it did him. The principle involved is the secret of a peculiarly successful form of respiratory gymnastics. The idea is to inhale freely and fully, but to exhale with effort and restraint, or with some obstruction to the out-going air. Thus the density of the expansive air within the thorax is so much increased that it will work its way into areas of lung where otherwise some air cells would be unused but for this prolonged effort.

Again, some congestion or infiltration might still

remain in out-of-the-way lung spaces were it not for the *push* given to the blood circulation by the increased pressure of this pulmonary air, which (during this forced expiratory effort) is in marked contrast to the rarefaction existing there during inspiration. This process of breathing is much like the natural stimulus of altitude. If assiduously practiced, it is an invaluable exercise.

It has been my own effort and purpose to make this principle adjustable to the needs of different invalids, and combine it with an inhalation chamber in convenient size and form to carry in the vest pocket and thus be always ready for use. The result of considerable experimenting is the Inhaler and Exhaler here illustrated. The first cut gives the actual size and form of the instrument as made wholly of hard rubber, and the second the interior mechanism.

DENVER SURGICAL INST. CO.

THE INHALER AND EXHALER.

The control of the medicated air inhaled is illustrated as follows (see cut) :

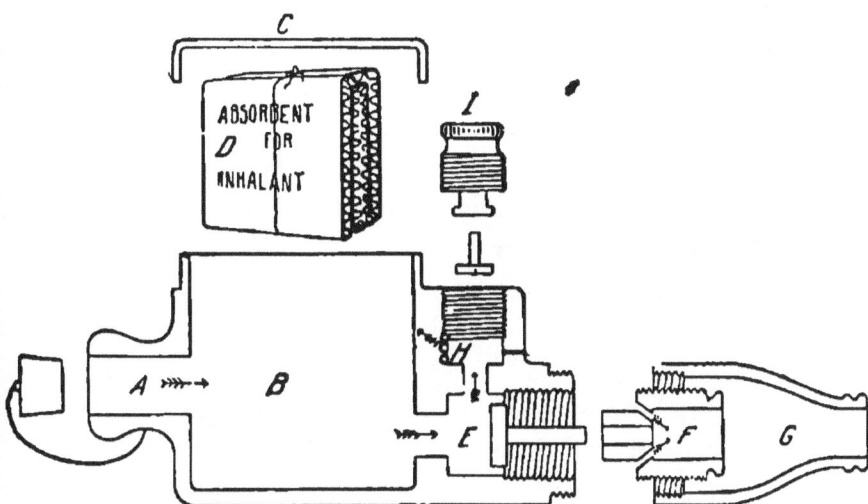

The air enters the Inhaler through nozzle "A," which is purposely made for the possible attachment of a rubber tube connecting with an oxygen tank, any desired vaporizer or gas generator. In the box "B" covered by cap "C" is the Absorbent "D," made of corrugated blotting paper a little smaller in size than the inner calibre of the box. This is intended to hold and gradually disseminate the various combinations of germicidal and healing oils, or other evaporating substances which may be used in this process.

The inlet valve " E," governed by screw head " F," may be reached by unscrewing the mouth piece " G " from the instrument. The inlet valve closes with expiration so that the inhaled air has to go off through the exit valve "H," which in turn is made adjustable by screw top "I." This constitutes the most important feature of this device. By the control of the exit valve, from 1-10 to 6-10 open, the tension of the air in the lungs may be increased at will from easy to difficult according to the need and ability of the user.

In using, the valves are supposed to have been adjusted to suit a given individual's strength (or as usually used, inlet valve " E " three-quarters open and outlet valve " H " three-quarters shut), the plug from nozzle "A" removed, and the absorbent " D " in box " B " semi-saturated with any desired inhalant. If the inhalant accompanying the instrument is used (specially designed for chronic lung and throat cases, perhaps tubercular, and composed of Pinol, Menthol, Phenol, Formaline, etc.) a partial saturation only once in four to six days will answer. The proper way to use the Inhaler is to DRAW LONG FULL BREATHS through the mouth-piece "G," the chest meanwhile being expanded and shoulders thrown well back. Then, without removing the instrument, EXHALE INTO IT WITH FORCE so that the entrance valve " E " closes with expiration, and so continue to EXHALE till all the air possible is removed from the lungs.

It is intended that this kind of use shall be forcible enough to redden the face of the user, and constant enough (say five minutes out of most of the waking hours) to keep open those air cells which through weakness or disease have a tendency to close, and to keep the affected lung area as nearly aseptic as possible. Besides, it is evident that such use favors hæmatosis (arterial blood making) by driving the oxygen in on the blood. The general results, particularly the gain in pulmonary ventilation as shown by increase of the spirometer and manometer records of the lung capacity and strength of an invalid, prove this to be one of the very best forms of pulmonary exercise, especially where bacillary complications are feared or known to exist. I believe that this form of exercise, with healing vapors thus introduced to the pulmonary seat of disease, is not to be compared with trying to reach the mischief by dosing the stomach. It is hoped that many young men and women may thus avert tubercular disease, especially in the lungs, and that this method of exercise, so similar in its results to those of the high altitude cure of consumption, may be made available to many in the lowlands who are unable to profit by the rarefied air cure.

Though this *Inhaler and Exhaler* has been in use only a short time, several hundred invalids could, if desired, bear witness to its grateful effects and beneficial results in their own cases. Having been completed since this paper was first written, it is introduced in this revision as made and for sale by the Denver Surgical Instrument Company, Denver, Colorado. Other and ordinary methods of inhaling in vogue are most of them opposed to the principle of distension of the lungs here advocated. The inspiration through a medicated sponge, muslin or packed gauze device, obstructing the act like nasal obstruction, has a tendency to draw together the air cells, and there is no after exercise effect to counteract this contracting influence in the lungs. Consequently, while the affected lung seems to be freed of catarrhal secretions by these faulty inhaling methods, the diseased parts are left in a shrunken state, i. e. worse than before. In obstructed fibroid and catarrhal lung affections the inhalation should be *free* and the exhalation intensified by increased air pressure within the chest.

Of course the opposite condition to this atalectic or fibroid shrinkage state is sometimes, though not nearly so often, found, namely, the over-distension of the lungs with air as in emphysema and bronchiectasis (dilatation of the air cells and bronchial tubes). For this condition of general emphysema I have devised a form of exercise which answers well its purpose to remove the stagnant pulmonary air and retained secretions and promote better circulation of both air and blood in the lungs. It is called " *The Chest Exerciser, or Emphysema Jacket,*" and is here briefly described.

It consists of a corset made to buckle or lace to

the shape of the chest in front, while behind it is left
open and each half is fastened by straps to rollers

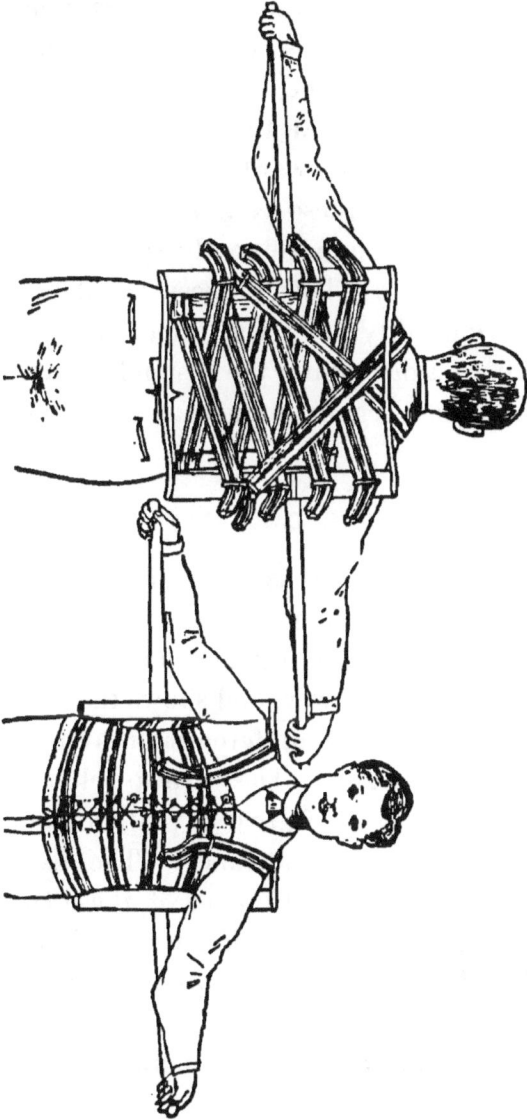

on the opposite side of the body, which rollers, by attached levers, are worked by the arms of the user in consonance with inspiration and expiration. During inspiration the arms and the levers are moved back, relaxing the corset, but during expiration the thorax can be squeezed to any desired extent by carrying the levers forward. Expiration, which is mainly at fault in these conditions, is thus chiefly influenced and made more nearly complete and sufficient than before. This process is intended to be used intermittently, say for fifteen minutes three or four times a day, until the distended air cells and bronchial tubes regain somewhat their normal resiliency.

LUNG VENTILATION, CAPACITY, AND STRENGTH.

How do we know there is a lack of ventilation of the blood and of air in the lungs? The foregoing reference to the fibroid or contracting tendency in affected lung tissue furnishes an answer, and shows the great utility of recording the semicircumferential movements of the two sides of the chest. They are so seldom alike in lung disease that it is of great advantage in diagnosis to note that rare circumstance. The explanation comes by noticing the difference of these two movements under forced inspiration and expiration when comparing the records of the *Spirometer* and *Manometer*. These instruments respectively tell, for the first named, the capacity in cubic inches of air

exhaled, and for the second, the strength of an indi
vidual's pulmonary organs by the pounds pressure
or millimeters of mercury force (*mm.*) shown in an
extreme expulsive effort. My aim has been to bring
these instruments to points of such excellence and
cheapness that every one, layman and physician alike,
can use them. The accompanying cuts (pages 25 and
26) are fair illustrations of the results of efforts made.

THE SPIROMETER.

STANDARD VITAL CAPACITY TABLE.

Computed from 5,000 observations (Hutchinson) of healthy persons, standing, while making a full expiration into the Spirometer after a complete inspiration.						Gymnasium Record of 2230 Amherst College Students (Males).
Height.	Males.	Females.	Height.	Males.	Females.	
4 ft. 7 in.	126	88	5 ft. 5 in.	206	168	194
4 9	142	104	5 6	214	176	210
4 11	158	120	5 7	222	184	224
5 0	166	128	5 8	230	192	240
5 1	174	136	5 9	238	200	256
5 2	182	144	5 10	246	208	270
5 3	190	152	5 11	254	216	286
5 4	198	160	6 0	262	224	302

As adjuncts to respiratory gymnastics they are of
great value, the manometer being a means of lung
development of no small account. Its habitual use
causes a steady and positive improvement in the
strength of the lung tissue. Emphysematous persons,
much affected, should not use it save for purposes
of diagnosis.

1. A positive defect in lung condition may be shown

because the Spirometer and Manometer each record a marked deficiency for the height and sex of a given individual, women recording usually 20 to 25 per cent. less than men.

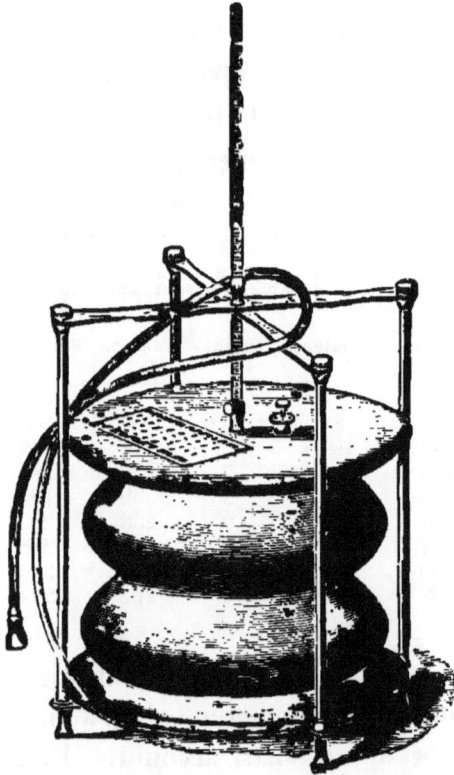

THE DENVER SURGICAL INSTRUMENT CO.

2. Again, in a given case the Spirometer may show a normal lung capacity, while the Manometer record is so markedly deficient (40 to 50 per cent.) that not only is insufficient lung strength inferred but

a reasonable suspicion perhaps aroused of some local disease either in the heart or lungs.

3. Again, the Manometer record may be good or show an excess of the normal force, while the Spirometer record is less than one-half what it should be for a healthy person. This is probably

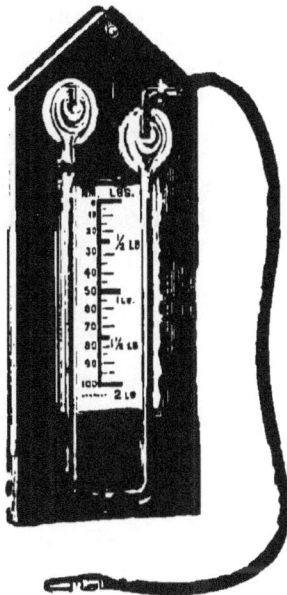

THE MANOMETER.

because of pleuritic or one-sided fibroid shrinkage, the result of disease. The inference is that the lungs have in a measure healed and gained the strength manifested even in their closed up condition. This is a compromise between health and disease, though nearer the former than the latter, with which exercise has very much to do. In fact, there would be no such arrest with the re-establishment of vigor except through the agency of respiratory gymnastics. Therefore the means of measuring and judging of one's respiratory capacity and strength are of great importance. If much deficiency or inequality of movement of the two lungs is found, it is much better for a young man to go to a good diagnostician and have his respiratory organs overhauled than to wait until he has to do so because of actual or progressive disease. Too often it happens that the lassitude and inability to exercise which is incident to disease, becomes constitutional, so to speak, and a habit of laziness gives feeble will-power to carry out any system of healthful respiratory exercise whatever. We thus come to recognize the great need of some incentive, either of pleasure or duty, to the end that the whole of the respiratory organs may be systematically used.

The writer once, on a declamation day, received the characteristic commendation of that most excellent college president, the late Mark Hopkins of Williams : "The young man has a good conception of what he wants to do." I wish I could impart a "good conception of what he wants to do" in the way of exercise to each one of the thousands of young men and young women whose sallow complexions, feeble circulation, short breathings, round shoulders, and flat chests betoken the depraved blood state which is already marking some portion of their lungs as the seat of the future bacillary battle ground. Let me make it a personal matter with each one who, because of past, existing, or approaching respiratory disease, needs light or medium gymnastics as opposed to heavy or severe gymnasium work. The object sought is a man normal in all his physical make up, and not individual feats of dexterity or muscular strength. For such a one I will formulate certain rules and forms of exercise best adapted for prevention of chest weakness, or for a chronic invalid's or convalescent's needs. These are not claimed to be exhaustive, but good at-home substitutes for more elaborate systems or for out-of-door activities.

Of course daily constitutional walks, hill climbing, horseback riding (with which for good effects bicycling, though excellent, can hardly compete because of the usual stooping attitude of the rider), tennis, ball playing, rowing, hunting, and fishing, are all most excellent forms of exercise, to be preferred by those for whom they are severally suited and safe because of the interest excited and resulting mental relaxation. But for the systematic home building up and strengthening of the weakened respiratory apparatus, these rules are submitted for individual practice with this distinct proviso : that doubtful or unsuitable conditions, feverish or irritable cases are always to be referred to the patient's physician for his choice of procedure. In fact, these forms of exercise are purposely graduated to enable the attending physician to determine how far a given person should proceed in a given time.

RULE I. *Cultivate regularity* in the care of your body; regularity, without so much precision as to be tiresome, in eating, sleeping, exercise, bathing, and the daily movement of the bowels.* This for some people is a cardinal requisite of good health. Let it be remembered that "procrastination is the thief" of vigor and vitality as well as " of time ; " and the languor and indigestion, with or without constipation of the bowels, which are the usual precursors of chronic pulmonary ills, favor mental irresolution and irresponsibility which are in no small degree to be overcome by *regularity*.

RULE II. *Look after the condition of your body's surface*, see that the skin is clean and therefore active, perspiration normal, and hands and feet warm. The morning *rub down* is a good thing. To those, and they are not few, to whom daily plunge or tub baths with soap are unsuited because of too great abstraction of bodily heat and lessening of the protective influence of a naturally oily skin, the *rub down*

* When the bowels are torpid, and hands and feet cold, the sipping of a glass of water as hot as can be swallowed an hour before breakfast is of much benefit; and so, especially if any sign of rectal weakness exists, are profuse flushings of the lower bowel with hot water from one to three times a week.

I have usually directed patients on retiring at night to take these enemata lying down, from a fountain syringe, and to hold from one to three quarts of the hot water in the bowel for twenty minutes if possible.

is particularly useful. It is all right for cleanliness, but its chief object is rather the resulting *reaction*. The warmth, glow, and exhilaration of this reaction after a bath is the criterion as to its length and coldness. To some, the most feeble, it may be enough to bathe above the waist in the morning on rising and below the waist on retiring at night, and to splash the surface quickly with tepid or cold water with the hand or wet towel for half a minute and then vigorously rub with a crash towel for three to five minutes. Others, not so delicate, can stand in a bath tub and squeeze a large sponge dipped in cold water on the back, and, after quickly going over the body, give themselves the towel rubbing. This is an excellent thing for anyone to do, even if perspiring from exercise, provided the resulting reaction comes quickly as it almost always will. To rub the body with cocoanut oil after a bath is an excellent procedure, especially in dry, cool climates.

Rule III. *Live as much as possible in the open air.* Let your exercise be carried on there, or with windows open if in the house. The more outdoor activity you have, the less gymnastic work will be required. Get as much sleep as possible (nine or more hours) during the night time in a well ventilated room; and a half hour nap at noon is also excellent. Do not worry but take life easy, (as the Irishman has it, " If ye can't take it aisy, take it as aisy as ye can.")

RULE IV. *Think about your chest position many times a day.* Whether sitting, standing, walking, or riding, *get into position.* This requires thought and will, till correct breathing becomes automatic. The

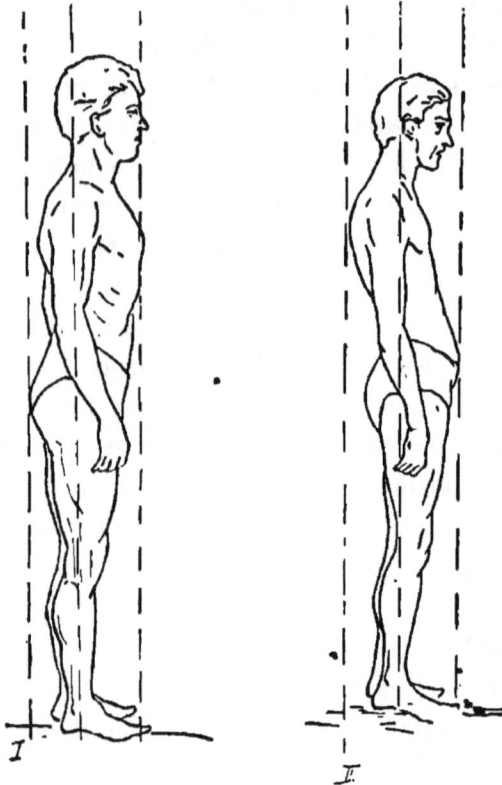

accompanying cuts from Checkley show the correct (I) and incorrect (II) chest positions.

The correct position is, head *up* and chin *in;* chest *expanded* front and shoulders *back* and *down,* the neck being *back* far enough to press against the

collar. If you stand in this way when your upper
garments are being fitted, your clothes will not be a
hindrance to right respiration as sometimes happens.
In this correct position, frequently practice breathing
by long continued inspirations; as you draw in your
abdomen, swell out the sides of your chest and pro-
trude your sternum (breast bone). Night and morning, while your chest is thus inflated, practice briskly
rubbing the chest from the sternum backward with
the palms of the hands.

RULE V. *Do not let the conventionalities of society
prevent your free and natural respiratory movements.*
The clothes should not constrict the chest or any
part of the body. This advice is of most value to
the ladies. Ordinarily, a woman with her corsets on
cannot properly perform the valuable exercises here
described. She is *usually* deceived as to the amount
of pressure exerted by these *vices*, so evenly adjusted
to her pliable frame. She is *positively* deceived as
to the support the corset is supposed to give toward
holding up the trunk of the body. It is simply the
acquired weakness of back and side muscles, caused
probably by long use of the corset, which she feels
it relieves. Had she properly used these muscles,
the delusion would have been unnecessary (?).

The accompanying cuts show the natural and correct female form without corset (III) and the de-
formed chest with corset (IV). It takes a great
deal of effort and self-control on the part of a young

woman to prevent the corset having this effect.
Some women have had the good sense to put two
towels lengthwise underneath their corsets when their
dresses were being fitted, so as to providentially
save for themselves so much more breathing space.
Doctors have preached, though unsuccessfully, about

IV

III

the constriction and crowding upward of the liver
and stomach, and downward of the abdominal and
pelvic organs, due to tight lacing; but, with reference to exercise, the chief effect to be lamented
is the unnatural variation or partial annulment of
respiration. The summit respiration is exaggerated
and the inferior costal or diaphragmatic movement

prevented. There is a wonderful amount of good in the systematic practice of these exercises to those women, especially the younger ones, who perchance appreciate the import of the foregoing and are willing to substitute a loose fitting corset waist for the usual corset ribbed with steel, whalebone, or their equivalents. Whether for males or females, the garments covering the trunk of the body should be loose fitting, and the covering next the skin is best of genuine wool of weight and texture to suit the season and individual needs.

RULE VI. *Practice front arm exercises and respirations combined*, with vigor and effort according to your strength and ability for the space of three to ten minutes three to six times a day. The accompanying cut illustrates one of these — the third.

First: Stand in correct position (Rule IV), fully inflate the lungs as you slowly raise the arms from the sides to the vertical, touching thumbs or backs of hands over head, and exhale as the arms descend.

Second: Draw back arms from front horizontal, with palms up, until the elbows are as far back past the sides of the body as possible, the elbows being kept close to the body. Inflate as you draw arms back, exhale as you return to position. Repeat this movement four to ten times. Last time, when elbows are back, slap chest lightly and quickly fifteen to twenty times.

Third: (See cut V) *Get into position.* Slowly raise arms from side forward, fingers straight out, till palms meet in front of forehead, fully inflating

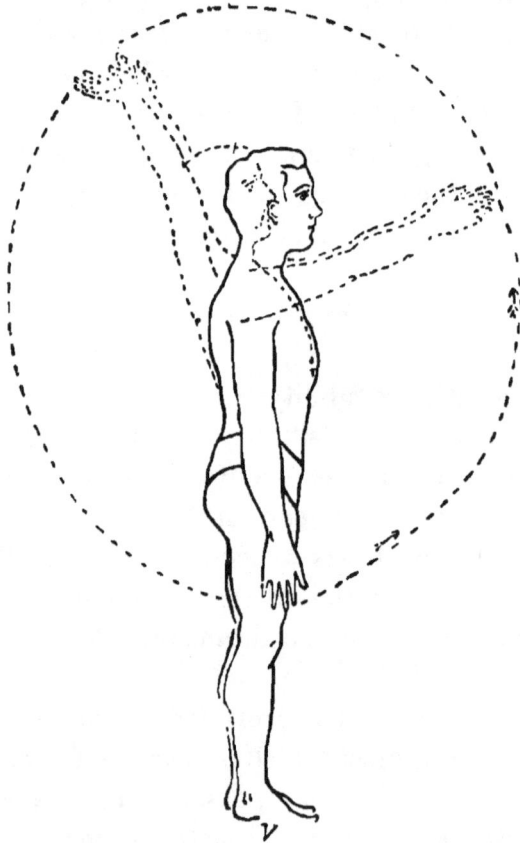

V

the lungs all the while. Then hold the breath until the largest possible circles are completed by both arms moving symmetrically over backward to position. Then exhale and repeat four to ten times.

An excellent variation of this movement is to step out diagonally to the right with the right foot as you draw in your breath, and swing up your arms front over the head as far as possible. Then alter-

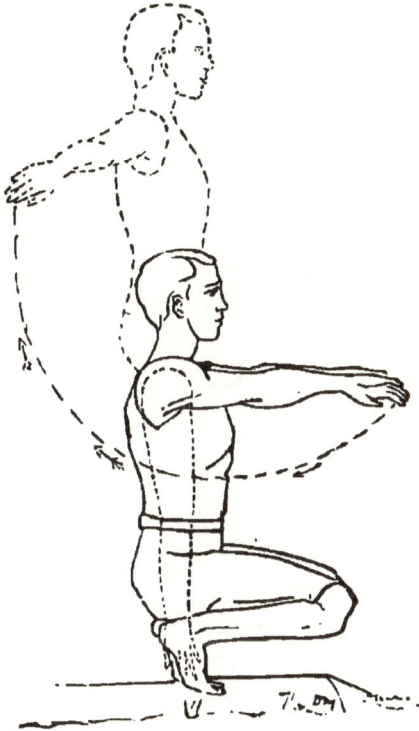

THE DRY SWIM.

nate several times by doing same to the left also, each time coming back to *position* by a spring with the advanced leg.

Fourth: "*The dry swim.*" *Position*, arms straight down at sides; exhale all possible as you go down

to floor. Then, resting on toes, slowly inhale as you
rise to position(see cut VI) making large swimming
circles, with arms stiff and palms out. Repeat four
to ten times.

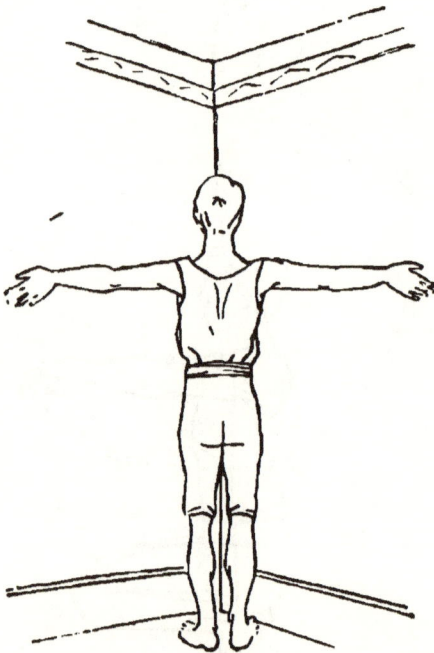

VII.

CORNER BREATHING EXERCISE.

RULE VII. *According to your ability* and free-
dom from any acute conditions, substitute or add to
the foregoing *the fixed chest breathings.*

A. Extend arms nearly to or above a horizontal at
shoulders and walk thus into the corner of the room;

then breathe four to ten full breaths. Do no more than you can accomplish without pain. The ability to get fully into the corner, with arms high, will come by practice.

B. Inhale as arms are extended laterally from cor-

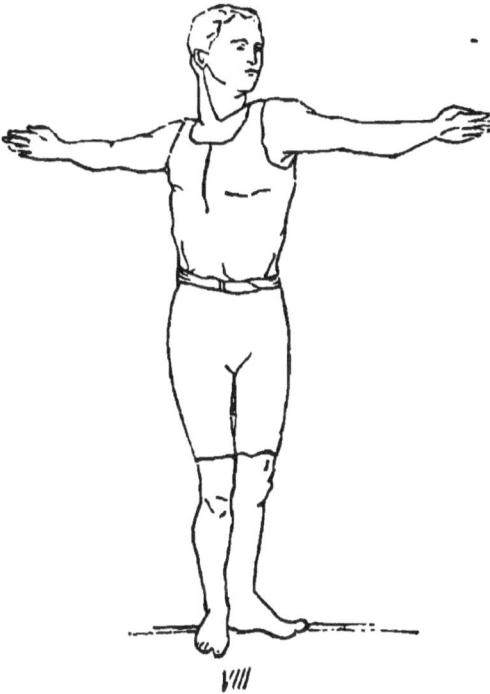

VIII

rect position, keep arms horizontal and face and feet to the front while arms and chest are swung quarter round; then with one stiff knee swing body forward until extended fingers nearly or quite touch the floor. Return to position and exhale. Take another breath and repeat with other arm forward. Alternate these movements four or ten times.

IX.

SWINGING CHEST EXERCISE.

C. Place two chairs about twenty inches apart and, with body stiff and straight, rest on toes and hands, as shown in cut. When down between the chairs take four to ten full inspirations and expirations. Raise the body from the down position to the extended arms, and repeat according to strength and ability to do so without too great fatigue.

If such severe exercise by practice becomes a pleasure and not a strain to a given individual, it is one of the strongest proofs possible of his power to resist

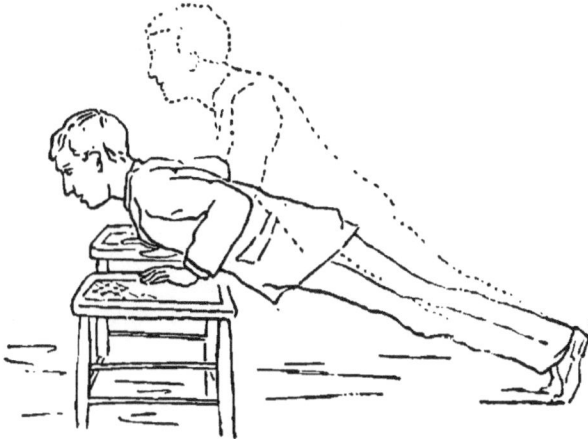

X.

DOWN BETWEEN CHAIRS.

pulmonary disease. They should be — the last especially — gradually reached by weeks' or months' practice of milder forms. If possible to combine them with the climbing of hills or frequent excursions to the mountains, the results will be so much the more salutary. As Professor R. J. Roberts, of Boston, says: "As a man breathes, so he lives. To half breathe is only to half live. So he must slowly and carefully develop his breathing powers."

FOOD

FOR

CHRONIC PULMONARY INVALIDS.

FOOD

CHRONIC PULMONARY INVALIDS.

Fothergill, in his "Manual of Dietetics," says:
"With modern knowledge of digestion and body
requirements, with a more perfect grip of the essen-
tial wants of the phthisical—of the importance of
arresting all outgoings, as well as that of the introduction of food at once digestible and nutritive—
the prospects of the phthisical are much brighter than
they were at the beginning of the present century."

Hygiene and Dietary, which are of so much importance to the human race, become laws of health,
the basis of happiness and prosperity, for the class
of people here intended to be reached. This class,
during some portions of their lives, constitutes a
larger part of the human family than is ordinarily
believed.

The vital statistics of "consumption" are very

Read before the Colorado State Medical Society, June 20th,
1894.

unreliable and deceptive in giving any accurate conception of the extent of the disease ; that is, if the blood condition tuberculosis is meant, as it usually is by the term consumption. It is assumed that it is recognized that many other diseases may also be induced by the blood perversions and impoverishments (as what we call anæmia, the diminution of hæmoglobin), which are simply symptomatic of a known or more often unrecognized tuberculosis. In mortuary records the deaths from pneumonia and pleurisy, brain, heart, liver, stomach and bowel diseases are usually otherwise accounted for than as due to constitutional conditions, and the arrested cases of tuberculosis are of course unrecognized. So when vital statistics attribute 13 to 14 per cent. of the world's mortality to consumption, it can reasonably be claimed that not half the number at some time affected with tuberculosis are represented in this estimate. Again, when a given section, as was the case in the State of Maine according to the census of 1880, gives 50 per cent. of the mortality, for those between the ages of twenty and forty, to consumption, then for adults of such age all other pestilences combined can be plainly understood to be of inferior importance to this one.

This question of anæmia, being closely allied with the tubercular process, is a vital one. Proof is not wanting that this condition exists even before the probably later evidence of tuberculosis, the bacillus, is microscopically demonstrated. There is undoubtedly too much procrastination, often on the part of the physician, from waiting to find some more positive evidence, as the bacillus of tuberculosis in the patient's sputum, before the recognition of the real danger. The important fact of the blood state, with other evidences of the lurking enemy, needs to be investigated. Dr. Rachford of Newport, Kentucky, in the November, 1892, number of the *Archives of Pediatrics*, presented valuable conclusions on the "Anæmia of Tuberculosis." These were based on 166 blood examinations of convent girls and seemed to clearly establish the following facts : *First*, That there is a relationship between "family tuberculosis" (inherited tendency) and anæmia. *Second*, That of those affected, the glandular cases (the involvment of the lymphatics indicating a much worse blood condition than when the disease is limited to the lungs) gave a largely decreased amount of hæmoglobin in the blood, *i.e.* less than did the purely lung cases, though in those the diminution was considerable. *Third*, That the anæmia of the pre-tubercular condition is probably due to "a secret or hidden tuberculosis of the deep lymphatics or other blood forming organs"; and *Fourth*, and finally, that "pronounced anæmia, without apparent cause, is strongly suggestive of concealed tuberculosis."

This is not the place to present other proofs that the beginning of tuberculosis does not date from the

46

discovery or known existence of the bacillus of that disease. There is always a previous susceptibility, caused probably by a vitiated state of the body fluids which produces a favorable soil or climate for the growth of the bacillus in man. It is a delusion to ignore the blood state (even though we cannot explain it), which must precede the discovery, not to mention the existence of the germ. The whole human race would perish of tuberculosis were it not for the resistance which healthy blood opposes to it. It is the *living cell* which must be depended on to maintain the supremacy and integrity of the human body. Were it possible to saturate that body with germicidal substances to keep away disease, which is extremely doubtful, it would produce, to say the least, a questionable state of health.

No, the life (health) is in the blood and the blood depends for its perfection on the proper selection and assimilation of food. Selection is necessary because, in defective states of the system, some foods are unassimilable and assimilation must be secured because, otherwise, what is eaten becomes a source of poison to the blood.

Again, without a healthful and bountiful supply of oxygen to the blood, the diseased conditions resulting from this defective or deficient assimilation would be always disastrous. But thanks to kind Nature, her provisions for every warfare are ample if we only have the intelligence to appreciate them.*

Say it is admitted that the perfection of the blood depends upon the proper assimilation of food, then a pertinent inquiry to be answered is, what relation does indigestion bear to consumption? Extended experience seems to confirm the statement that either catarrh of the stomach and intestines or dyspepsia are very often precursors of chronic pulmonary ills. This being the case, how can they be explained and thus prevented?

A good understanding of the physiological facts pertaining to normal digestion is necessary to the correct appreciation of those pathological states which invariably accompany abnormal digestion. A great deal happens in the stomach that this patient and much abused organ does not complain of, and so we are mentally unconscious of our perfect or imperfect digestion. "The healthy stomach, when empty, is contracted and its surface is pale; its vessels small and tortuous. When, however, food passes down the œsophagus and is received into the stomach, the pink, velvety appearance of the mucus membrane is seen to give place to a brighter, slightly darker shade; the vessels dilate and become more

---

* For further points on this matter, reference is made to the preceding essay, "Exercise for Pulmonary Invalids," the burden of which is, the thorough oxygenation of the blood as a preventive of disease.

full of blood, and the secretion of gastric juice commences.

Experiments have further shown that the same results follow upon the introduction of food through an external opening, or upon stimulation by means of a smooth body introduced from without and gently rubbed against the inner surface of the stomach. If, however, the stimulus be carried to such an extent as to cause irritation, effects exactly the opposite of those described above are produced — namely, contraction of the vessels, suppression of the gastric juice, and the secretion in its place of a quantity of mucus."*

This physiological process is of great import in our present inquiry because it is undoubtedly in the stomach that that "irritability" commences which afterward, farther along in the alimentary canal, through malassimilation, helps to complete the condition of dyspepsia. The irritability is there whether the cause in the stomach be bulky or improper food, defective or deficient gastric juice, innervation, or excessive fermentation producing gas or other chemical action.

The question, how impoverished states of the blood may favor this condition of irritability, is not so very difficult to explain to one who understands the circulation and the varying effects of changed or

* "Food and Dietetics," by R. W. Burnet, M.D., Vol. XI, Wood's Medical and Surgical Monographs.

deficient blood tension or supply in the internal
organs. But even to the uninitiated layman it is possible to explain the converse of the above, *i. e.*
how this irritability and inefficiency of action represented by dyspepsia may lead to lung disease and
so often be a cause of it. Undoubtedly the two
go hand in hand and are interdependent; but indigestion once inaugurated and afterward fostered
by sins of commission in eating and drinking must
lead through the defective assimilation of food
ingested to a defective supply and quality of blood
sent to all important organs. The defect in the
blood will be noticed in the lungs particularly, where
the vulnerability to atmospheric influences will be
manifested. This dyspepsia is not usually a cyclone,
but a steady wind; a process which, through the
resulting lack of proper nutrition, makes itself felt
in all parts of the body, and the sufferer is consequently constantly retrograding in physical stamina.
Meantime the malproducts, developed by the decom-
position of ingested foods which have not under-
gone proper conversion into assimilable forms because
of the indigestion, are practically poisoning the whole
body through their absorption into the general
system.*

* If to any one, probably to Dr. J. H. Salisbury of New York
the credit is due for announcing that the chief malproduct
which is to blame for all this evil in the blood is an acetic acid
fermentation. Later Dr. Ephraim Cutter, in his "Clinical Mor-
phologies," nicely demonstrates to his own satisfaction by mi-

The acid and fermenting states of the blood induced, the torpid peripheral circulation, and the
crowding of the congested blood to the overburdened
internal organs, seem to be correlated. The usually
defective skin elimination and the deficient respiratory activity are also noticed in these pretubercular
subjects. That condition of the body's circulating
medium has been reached and is maintained in the
lungs, which is most favorable for infection. Hence,
by preference, the pulmonary tissue presents the

crophotographs of high magnifying power taken from studies
of tuberculous blood, that the spores of "vinegar yeast" exist
in the blood plasma and in the white corpuscles, which corpuscles we know are in excess of the red in anæmia and leucocythemia. The claim of Dr. Cutter to this discovery is worthy of
the most careful study that histologists can bestow upon it, to the
end that its accuracy may either be proven or disproven. My
own study thus far tends to confirm the views announced of
the existence of a ferment in the blood in some due proportion
to its depraved condition, which condition underlies scrofula,
hip-joint disease, chronic rheumatism, chronic dyspepsia, etc.,
as well as tuberculosis. We are waiting for some physician of
recognized authority in bacteriological research to give us the
much to be desired thorough elucidation of the various
clinical indications to be drawn from the microscopic investiga-
tions of the blood in such wasting diseases, and particularly in
consumption. Until then we can only conjecture regarding this
excellent diagnostic means of the pretubercular state.

To the same discriminating judgment also may be submitted
the late claim of Dr. R. L. Watkins, that he has found the
source of tubercular infection in the previous existence in the
blood of "*the third blood corpuscle.*" To verify the causative
relation of these so-called "corpuscles" to tuberculosis, he was
willing to have himself inoculated with bacilli tuberculosis

most convenient and congenial lodging-house for the germs which lurk nearly everywhere in the inhabited atmosphere. Then in the lungs we find the sequel of a process which dates back to malassimilation chiefly in the duodenum and the balance of the small intestine, previously induced by indigestion in the stomach. Great emphasis is given to these considerations by the rareness of recovery in cases of . tuberculosis where persistent gastric disturbances exist. The form of tuberculosis which seems so

which he did, his own blood, as he claimed, being "free from these corpuscles." The difference between this microscopic evidence of immunity and that of Salisbury and Cutter is not yet clearly shown. The whole extremely interesting and promising subject is in a stage of development waiting for the more conclusive report of that analytical and differential histologist (as yet unnamed) who should receive honor second only to Koch himself.

[At the same meeting of the Colorado State Medical Society at which this paper was read, the author introduced a resolution which was carried, offering a prize of one hundred dollars for the best essay on "*The Diagnosis of Tuberculosis by Microscopic Examination of the Blood.*"

A committee of award was appointed, composed of Drs. C. Denison, H. A. Lemen and S. E. Solly, to the first of whom all essays were to be sent by April, 1805.

A paper in the English language, condensed to read in thirty minutes time, giving preference to new evidence and the detection of the pretubercular state, is desired. All stages, however, to be microscopically differentiated, especially from simple anæmia and leucocythemia; the committee to reserve the award for an essay they deem sufficiently meritorious, *i. e.* the rules to be observed enabling a diagnosis to be made from the blood alone without the patient being seen.]

evidently to start in dyspepsia and malnutrition is
most serious, and its symptoms, the loathing of food
of any kind and especially of fats and nitrogenous
foods which are most needed, show the importance
of first attending to the alimentary canal.

If, in discussing this question of "foods" in consumption, we are compelled to take sides with or
against the vegetarians, then I am ready to say that
a considerable experience leads me to favor a mainly
nitrogenous or meat diet as preferable for most invalids. The arguments in favor of animal diets are
based,   .

1st. Upon the animal origin and constituents of
milk.

2d. The carnivorous nature of human teeth. This
may indicate a mixed diet, but twenty out of the thirty-
two teeth of the adult man being carnivorous in
type, two-thirds of his diet would seem to be indicated as more appropriately carnivorous or meat.

Whole wheat is rich in vegetable albumen, gluten
(10 to 35 per cent.), phosphate of lime and fibrin,
with a form of starch easily convertible into fat and
heat. This, then, is the best form of food, properly
prepared, for the other one-third of the human diet.
Starch in the whole wheat is not in the excessive
proportion found in other cereals. In such wheat it
is .57; in ryè, .64; oats, .65; barley, .66; corn,
.67; rice, .88; while in white flour it is .75. How,
it may be asked, has the flour gained its 18 per

cent. more starch in the process of milling? In Johnson's "How Crops Grow" we find that in 1,000 parts of substance wheat has an ash of 17.7 parts, flour, 4. parts, a loss of over three-fourths. Wheat has 8.2 parts phosphoric acid, flour only 2.1 parts. Wheat has 0.6 lime and 0.6 soda, flour only 0.1 of each. Further, wheat has 1.5 sulphur, 0.5 sulphuric acid, and 0.3 silica, while flour has none of these. Really it seems as though the refinements and thoroughness of milling have been of great damage to the human race by taking from wheat food (flour) these valuable salts.

3d. Man's stomach is not like that of the herbivorous but of the carnivorous animal. In structure and function it is designed for the digestion of lean meat. It is suited to the disintegration of muscular fibre and the dissolving out of musculine and albumen fitted for absorption, in which preparation the gastric juice plays an important part. This is the commencement of the process which is completed farther along through the work of secretions from the liver and pancreas, and by the intestinal juices.

4th. It is claimed that the constituents or qualities of which the human body is composed, are best supplied by that kind of food which contains the same ingredients, i. e. water, musculine, albumen, lime, fibrin, and fat, in the nearest to an assimilable form.

5th. The natural craving of human beings for

lean meat is a strong indication of its need. This has always been marked, and a large proportion of human beings have always been known to live on meat.

It is claimed that lean meat, properly cooked, is prepared in the stomach for digestion and assimilation sooner than starch food.

To judge what foods are best, the stomach must be in a normal state, i.e. free from acids, ferments, and catarrh or excessive mucus. Drinking hot water usually accomplishes this desirable end, cleanses the stomach, and favors healthy action of skin, bowels, liver, and kidneys. If objection to hot water is insurmountable, weak tea or weak bouillon may be substituted. At the same time the rules for exercise, bathing, mental diversion, sleep, etc., should be obeyed, and the bad results of too much and too fast eating and an unsuitable diet, should be avoided. Assimilation, not quantity, is the guide. Lean beef is the tissue builder, par excellence; and the highest muscular development and greatest powers of endurance are always sought, in training athletes, through the agency of the lean beef diet.

On the other hand, starchy food and sweets tend to produce acidity, fermentation, intestinal catarrh, anæmia, and favor a condition of blood more conducive to consumption. The use of the lean meat diet in disease is based upon the fact claimed, that it produces the maximum amount of nutriment

with a minimum of digestive effort. It is claimed
that microscopic investigation of the blood under a
lean meat diet demonstrates the blood-making tendency of the same. The symptoms which sugar and
starch, predominating in the diet of little children,
produce, are those so often associated with scrofula
and consumption, i.e. catarrhal tendencies, skin and
glandular affections, capricious appetite, and catarrhal and acid conditions of the bowels, protuberant
abdomen, and deficient muscular tone. These tendencies are successfully averted by proper and systematic feeding of children on milk, eggs, beef, and
whole wheat flour or coarse-grained cereals.

It is claimed that an excess of starch and sugar
in the diet favors gastro-intestinal catarrh, which
disease in turn is best influenced by the beef and hot
water regimen.

Defective digestion, whether from physical ills or
unsuitable nutriment, leads to or goes with abnormal
blood and with malnutrition, which, in turn, account for the failure of vital action, and of the
organic tissues to sustain a healthy relation to the
human economy. Hence the foundation of consumption may and probably does lie in this defective
digestion at the start.

On the contrary, it is claimed by the opponents
of the lean meat diet, that it is too heating to the
blood, leads to dyspepsia, rheumatism, Bright's
disease, gout, etc., causes tape worm, increases urea

in the urine, and makes man combative and passionate.

·Well, as to the heating proclivities of the two diets, meat does not contain the carbon in order to be heating, while starch and sugar (represented in oatmeal, potatoes, rice, white bread, and some fruits), not only are fat producing as carbohydrates, but they undergo fermentation and are then temporarily very heating. If, by what is termed "heating of the blood," that improved circulation is meant which is indicated by the evenly warm hands and feet, seemingly produced by the exclusive hot water and pulp of meat diet in some dyspeptic cases, why then that is the kind of blood heat that is wanted. The mechanical effect of the hot water, taken in conjunction with the lean meat regimen, is very important as improving the solubility and flavor of the meat, and at the same time quickening absorption as well as elimination.

The causing of tape worm is indeed infrequent and, if it does occur, must come from the lack of proper cooking of meat. An excess of urea is not dangerous, an increased amount being thrown off by exercise as well as by animal food. As to Bright's disease, rheumatism, and gout, the assertion that animal diet causes them is not proved, and it is probable that other excesses, as stimulants, too high living and too much sweets, have more to do with the production of these diseases than does that kind of diet (beef juice, pulp of meat, underdone steak, etc.), which is so non-feverish and at the same time eminently restorative and upbuilding to the consumptive invalid.

A practical application of the principles of dieting to the individual's needs is of great importance, whatever system of feeding is decided upon. The usual hap-hazard, experimental method is founded on ignorance, if it has any foundation, and is not a success by any means. There is not time in this paper to present and discuss all the claims of the great variety of specially prepared forms of food on the market. They are largely for use in the feeding of infants. There are some of them, however, which we may select as in addition suitable for grown people, for in many instances of adult indigestion we have to consider such an extreme of enfeeblement, or such an entire cessation of the functions of digestion and assimilation, that it is the best plan to give up all solid foods and go back to first principles; in other words, it is preferable to commence over again on the basis of infant foods, which are more or less prepared for immediate absorption. This consideration shall determine the policy to be pursued in this practical presentation of our subject.

We will commence with the feeblest and most serious class of cases, and work back towards perfect digestion as a basis of health, which is the good we are seeking.

This reversal of the natural order of sequence of disease suggests the following arbitrary classification : —

1st. *The terminal dyspepsia of tuberculosis.*

2d. *The initial dyspepsia of tuberculosis.*

3d. *The preventive dietary of the supposedly healthy,* or if preferred the preventive dietary of the possibly infected.

First, I have already referred to a very unpromising class of cases characterized by persistent gastric disturbance. There comes a time, which is earlier or later with different patients and sometimes seemingly dependent upon the acuteness of the attack, when the individual's resisting power is so feeble and the blood so charged with the tubercular virus, that apparently the nervous influence which governs the assimilation of food is overcome, and the digestive process is at a standstill. There is loathing of food, especially of meats and fats or oils, and an abnormal desire for cold drinks, starchy fruits, sweets and acids, which, through the increased fermentation produced, only add to the imperious fire that is slowly consuming the body.

These severer classes of cases are, or surely ought to be, under the strict advice of the physician. It is for his consideration and adoption that the following regimen is suggested as a sample of what ought to be done, and not as a complete *résumé* of all that can be done. A decided and radical change in diet should be made at once. The doubt about the digestion of the casein of milk may perhaps shut off from use cow's milk, ordinarily the most natural invalid's food.*

The extreme of caution, likewise, dictates either (1) a predigested liquid diet or (2) a wholly nitrogenous regimen, until it is certain that fats and starchy cereals can be well taken care of.

1. The predigested liquids are such as the Arlington Chemical Co.'s "Liquid Peptonoids," or Fairchild Bros. and Foster's "Panopepton," a prepared and digested extract of beef and bread. One of these two excellent preparations should be given according to the directions on the bottle and alternated every two hours during the day, and once each at night (or in all, say four times each in twenty-four hours), with Mellin's Food prepared at first, perhaps, without milk or with one-half the proportion directed on the package. Mellin's Food, a selected and predigested combination of wheat and barley, is

---

* Sometimes milk may be made to agree when otherwise it would not because of its constipating tendency or the inability of the patient to digest it, by combining it with Seltzer water in equal parts, by adding a little salt to the milk, taking some Hunyadi water with the morning portion, or by peptoniziug it. Though for some persons milk is a most appropriate constituent of every meal, there are many to whom its generous or even moderate use is not at all suited. I therefore insist that it must be first definitely known that milk agrees and is digested, does not leave a furred tongue and a torpid liver, before milk drinking is incorporated in an invalid's diet.

an excellent preparation in the form of a dry powder, in which digestion is carried so far (the starch being all converted into malt sugar—maltose—and dextrine) that fermentation in the stomach is not only prevented, but the after absorption is easily accomplished. Mellin's Food is thought to aid in the digestion of cow's milk with which it is usually given, and in which combination it contains all the food elements the human system requires. If it is too sweet in taste use proportionately more salt in preparation. I have seen many cases of dyspepsia recover on a diet of Mellin's Food with very little other food or medicine.

2. The desirable nitrogenous regimen in liquid form can best be obtained in the expressed juice of underdone broiled steak. The juicy, round steak is preferable, and it should be broiled over hot coals very lightly on the two sides, and then, if necessary, cut in small enough pieces to go into the receiver of a meat juice press. An Osborn No. 1 or Bartlett No. 5 can be obtained at the hardware store if a cheap machine is desired; but the one or two larger sizes give proportionately greater compressing power, all the way from 1,000 to 3,000 or more pounds, sufficient to expel all the nutriment there is in the meat. This expressed nutriment looks like blood,* and is chiefly in the form of albumen which must

* There is a notion abroad that drinking warm beef blood fresh from the slaughtered animal is of great benefit. I have

not be heated above 100 degrees F., for this coagulates and for the most part destroys it. This is what is the matter with over-broiled steak, and as for ordinary "beef tea" its use ought to be abandoned. A dessertspoonful to two tablespoonfuls of this juice, well salted, diluted with warm water as preferred and taken every two hours, will sustain life, being almost immediately absorbed and at the same time giving entire rest to the digestive function. For variety it can be alternated with Mellin's Food, or later on with other prepared foods, as Nestlé's Food, Ridge's Food, or Imperial Granum. The stricter diet thus laid down may be adhered to from one to four weeks, when, if the appetite and digestion warrant the change, the patient can be allowed to pass on to the diet specified for our second classification, *i. e. The initial dyspepsia of tuberculosis.*

As already imperfectly explained, the coincidence of dyspepsia with commencing, latent, or established tuberculosis, is so striking (whether the indigestion be the cause, as Dr. Salisbury claims, or the accompaniment, of the infection, as seems more reasonable to believe) that it is impossible to conceive of a more important subject for study than the relation of food to this disease.

never seen any good result, but on the contrary, have known several consumptives' stomachs upset by this practice. But the rectal injection of defibrinated blood, say two or three ounces twice a day, is strongly recommended by Dr. Andrew H. Smith of N.Y. and others.

It is unfortunate that previous and in a measure erroneous conceptions of tuberculosis have so imbued the writings of Dr. Salisbury, the father of the meat pulp and hot water diet theory ; yet it must be understood that Dr. Salisbury worked and wrote before the close association of a germ with tuberculosis was demonstrated by Professsor Koch. When in the dark about this scourge of mankind, it was rather creditable to the medical profession at large that they did not admit Dr. Salisbury's claim that he had found the cause for all diseases except those arising from "injuries, poisons and infections." He states in the preface of his book, "The Relation of Alimentation and Disease," "I started in without theories, without prejudices," yet his premise seems to have been that the abnormal states underlying consumption were "paralytic," whatever that means, and chemical effects seem to have been substituted for what we now know to be and to have been bacillary results.

There are other reasons why meat is pre-eminently the best diet in the dyspepsia of tuberculosis than that meat, because of the absence of starch and sugar, does not ferment, though that fact is of great value. It is the failure of the nervous force which has to be met by this nerve-energy-giving food, and the more we study consumption the more we must recognize that the wasting of nerve power is a most important factor in this process of decay. It is mainly from albuminoids that the system originates and is renewed, and albumen is not made by digestion but must exist already in the undigested food. Though albumen largely enters into the composition of other foods it is in the pulp of meat that it is found in its freest state. Besides, the power to assimilate other foods is much impaired or lost, and we should not forget what Bartholow says, that "it is not the quantity swallowed, but that digested and assimilated which contributes to the nourishment of the body."

Hence the meat pulp and hot water diet has proved to be very beneficial in the temporary or more persistent attacks of indigestion incident to incipient and even long established tuberculosis. This diet should be continued long enough, and not be too much modified by other foods to lose the desired result, namely, the perfect assimilation of what is taken.

Of course beefsteak can be scraped, which is laborious, or an expensive apparatus like Salisbury's for extracting the pulp can be used. A practical and economical way to get the pulp free from the fibre, tendons and fat, is to buy a medium sized Hale meat-grinder (size No. 2 will answer for a family of two to four persons). The meat is run through this grinder six to eight times, and each time what is not wanted catches on the knife between the revolving cylinders and is removed. Finally rather more than half of the weight of meat used is obtained as pulp. This can be suitably flavored, as with horseradish,

and eaten lightly cooked or nearly raw. A good
way is to press the pulp into cakes not too hard,
salt and pepper them to taste, and lightly and
quickly broil over hot coals so that the centre re-
mains red and juicy. This meat served in conjunction with or preceded by the sipping of a cup of hot
water, is easily dissolved and assimilated. Immediately, or after a variable time, according to the
severity of the case, this plan can be combined, or
perhaps better alternated, with the use of cereals;
as fresh Pettijohn's Breakfast Food, Imperial Granum, cracked wheat and cream, soft poached egg on
toast, etc. Or between the meals of the meat pulp
a raw egg dropped in a wine glass of dry sherry,
or the following combination of an egg and Mellin's
Food may be tried; one egg, two tablespoonfuls
sweet cream, two teaspoonfuls sugar, one tablespoonful Mellin's Food, and one-half pint milk.
Beat the egg well, add the cream and sugar and
beat again, then turn on to the Mellin's Food, previously dissolved in a little hot water, mix thoroughly, and add the milk. This may be salted or
otherwise flavored as desired. Another way to take
the pulp of meat is to season it, and spread it in
thin layers between thin slices of bread.

An excellent form of food to use separately or in
conjunction with the meat diet, is the recent inven-
tion of a Denver gentleman, namely, The Cereal
Machine Company's product of Shredded Whole

Wheat. It consists of the preserved ingredients of the whole wheat berry (none of it excluded by the usually overdone process of sifting), which, after a thorough boiling, is shredded by a machine specially designed for the purpose. This makes a light, fluffy material which is eaten with cream or variously cooked. The usual custom is to bake it in cakes, to be used as bread or warmed over and served with cream. Many articles of food, like pie crust, are made of this material, light and porous without baking-powder or shortening. If not consumed within forty-eight hours this food is roasted, after some peculiar fashion unknown to me, and made into a very palatable coffee. The advantage of this beverage is, that its nutrient elements are thus substituted for the stimulant effect of ordinary coffee or tea, while at the same time it is as good as the hot water, previously found to be so beneficial, and preferable to cold drinks taken with meals.

The use of stimulants is, to be sure, a mooted question with many well meaning and earnest people; but the unprejudiced physician sees a positive benefit result from a weak invalid's taking a glass of ale or porter twice a day, or from a sleepless one having a hot milk punch or whiskey toddy in the middle of the night.

Fothergill in his "Handbook of Treatment" strongly recommends the following, " rum and milk in the morning ere dressing." " Take a half pint of new milk, and add to it an egg, a teaspoonful of powdered sugar, some grated nutmeg, and one or two teaspoonfuls of good old Jamaica rum; stir all together well, and let it be taken by the patient in bed." "It is desirable," he says, "that a little further rest in bed should precede the process of dressing."

There are, to be sure, other forms of solid and liquid food which will, like the above, be good to vary the strict meat diet with as soon as the time shall have arrived that normal assimilation takes place, and the usually normal craving for other and more food is experienced. The chief consideration and care should be that these substitutes to be experimented with, should be easily digested, prepared in an appetizing way, and should not too largely represent the starch and sugar elements of food. For the same reason fruits, like bananas and melons, have a doubtful place with any meal, though oranges, apple sauce, pickles, nuts, and cheese may be sparingly allowed. Thus by degrees the six meals a day, mainly of nitrogenous food, may become merged into the three principal day meals usually taken by the healthy; the heartiest, dinner, coming by preference for invalids in the middle of the day, when possible. Even then the broiled pulp of meat will go well for breakfast, or breakfast and supper; and for something warm at bedtime a goblet of Mellin's Food, two tablespoonfuls of the dry powder to the goblet

of equal parts of hot water and milk, answers quite
well and conduces to a good night's rest, and is
suited to the feeble digestive powers, if, for reasons
just given, a cup of hot bouillon, a milk punch, or
sherry and egg is not preferred. Thus it is the object
of this method, in one week or one or more months,
to work up to the third classification of diet, namely :
The diet for the supposedly healthy. This diet, to
be preventive of tuberculosis where once it has
existed or become latent in the system, must be
sustaining to the powers of life to the end that a
vigorous warfare be kept up for the supremacy of
the living cell. While the nervous system is sus-
tained by a plentiful mixture of nitrogenous (meat)
and phosphatic (fish, oysters, etc.) food, the wasting incident to the dreaded disease must be prevented
by an abundance of the hydrocarbons, of which fats
and oils stand at the head. At the same time assimilation should be promoted and secured by something warm or warming with every meal, as chocolate, cereal food coffee, hot milk or water, ordinary
weak coffee or tea, or at dinner a single glass of
light wine.

Thinking it would be a good idea for the house-
wife or landlady to have a sample or samples of bills
of fare, I have selected the following as a good list
from which to choose one or more dishes for each
meal. It is not intended that this diet shall be the
limit of what kinds of food may be selected but only

suggestive. Seasons, latitude, markets, and last
but not least the purse, will dictate many changes
or modifications : —

Breakfast, 7.30 to 8 A.M.

Either cracked wheat and cream, Pettijohn's Breakfast Food, Cereal Machine Co.'s Shredded Wheat with
cream, cerealine, wheaten grits with cream, broiled
hominy or fried mush.

Either broiled steak, fat mutton chop, broiled fish,
particularly the white meated fish, well cured ham with
fat, breakfast bacon with poached eggs.

Either baked or creamed potatoes.

Either corn bread, brown bread, whole wheat bread,
corn meal griddle-cakes, Graham muffins.

Either "cereal coffee," cocoa or chocolate made with
milk, weak coffee with cream and hot milk, or if these
disagree, hot water. (If hot water has been taken one
hour before breakfast the result is beneficial in many
cases.)

If a lunch is needed at 10.30 A.M., some pulp of meat
spread between thin slices of bread, with hot water or
some broth, an egg-nog or sherry and egg will answer.

Dinner, 1 P.M.

Either raw oysters or clams. Soup—oyster, cream,
green pea, celery or tomato; bouillon, mutton or chicken
broths.

Either broiled or baked fish, fricasseed chicken, roast
beef, mutton or game. Baked or mashed potatoes, with
either string beans, green peas, spinach, squash, aspara-
gus, cauliflower, tomatoes, celery, lettuce. Bread—whole
wheat or brown.

Dessert : puddings—tapioca, Indian, corn starch, bread, apple-tapioca, chocolate, snow or steamed apple; wine jelly, boiled or baked custard, charlotte russe.

If lunch at 4 P.M., white wine, crackers and cheese, or milk and tongue sandwich; fruits.

Supper, 6 to 6.30 P.M.

Either oyster or fish soup, warmed over meat, broiled pulp of meat, milk toast, macaroni and cheese, toasted brown bread, Graham bread, or corn bread. Cereal food coffee or weak tea, hot milk, meat or fruit jellies, cheese.

If lunch before going to bed, 9.30 P.M., beef or clam bouillon, hot milk; a little wine, crackers and cheese, or a tablespoonful of Mellin's Food in a glass of milk; if no lunch, then bouillon or hot milk in the middle of the night may do well.

For some invalids who seem to need it, I have advised this full diet in combination with that designated for our first classification, substituting an ounce or two of the expressed juice of meat in place of the lunches named above.

If a person is habitually constipated, hot draughts of water night and morning, fruits (figs, etc.) and the coarser vegetables combined with the diet, are of service.

Reference must be made to the companion paper to this on "Exercise" for rules as to care of the body, regularity in habits, chest rubbings, forms of exercise, etc. It must also be explained that the foregoing regimen may be too largely composed of animal food for many healthy people; especially is this true for those advanced in life. Then there is naturally a falling off in the ability to take care of nitrogenous food, while at the same time the capability to obtain and the incentive to enjoy a hearty cuisine have increased. Vegetables and fruits seem to be more appropriate foods for these, while with very old people there is a natural tendency to return to the liquid diet of infancy.

Climate, seasons and occupations, too, make a great difference in the proper choice of foods. The indication for animal food (meats and oils) in cold weather is so much more imperative than in summer time that the warmest season of the year (notwithstanding the ability and incentive to be then much out of doors) is not so favorable for the kind of diet which is best for the consumptive. Here is a strong argument in favor of accepting cold as an indispensable element in the preferable climate for the phthisical invalid.*

For instance, a moderately healthy person, actively engaged in out-door pursuits in winter (or undergoing much mental strain as well as muscular waste) requires or does well with meat at every meal. Such

* For further corroboration of the importance of *cold* versus *warmth* and for facts bearing on climatic treatment, see the author's brochures, "The Preferable Climate for Consumption," and "The Rocky Mountain Health Resorts," published by Houghton, Mifflin & Co., Boston, Mass.

an one may in summer be well satisfied with the substitution of cereal foods and fruits and a very moderate use of meat say at one meal each day—if he is not required to or does not voluntarily perform active muscular exercise in such hot weather.

Apply the same rule to the consumptive's diet, and some, but a lesser, variation to fruits and starchy foods would seem to be advisable, because we can never lose sight of the greater proportionate waste of tissue and the poverty of blood in this class.

It has been said that "the majority yield to inclination rather than are guided by intelligence and judgment"; hence the great importance of this food selection question to the consumptively inclined whose inclinations are so often pulling against their better judgments. The author will feel amply repaid if the foregoing plan of battle, though imperfectly presented, shall aid invalids to conquer or keep in check this foe, tuberculosis, and the doubtfully affected to follow a line of duty which shall enable them to avoid the necessity of an actual contest.